Entrapment in Garage Kills One Firefighter

San Francisco, California

Investigated by: Scott M. Howell

This is Report 084 of the Major Fires Investigation Project conducted by Varley-Campbell and Associates, Inc. under contract EMW-94-C-4423 to the United States Fire Administration, Federal Emergency Management Agency.

Homeland Security

Department of Homeland Security
United States Fire Administration
National Fire Data Center

UNITED STATES FIRE ADMINISTRATION
FIRE INVESTIGATIONS PROGRAM

T he United States Fire Administration develops reports on selected major fires throughout the country. The fires usually involve multiple deaths or a large loss of property. But the primary criterion for deciding to do a report is whether it will result in significant lessons learned. In some cases these lessons bring to light new knowledge about fire--the effect of building construction or contents, human behavior in fire, etc. In other cases, the lessons are not new but are serious enough to highlight once again, with yet another fire tragedy report. In some cases, special reports are developed to discuss events, drills, or new technologies which are of interest to the fire service.

The reports are sent to fire magazines and are distributed at National and Regional fire meetings. The International Association of Fire Chiefs assists USFA in disseminating the findings throughout the fire service. On a continuing basis the reports are available on request from USFA. Announcements of their availability are published widely in fire journals and newsletters.

This body of work provides detailed information on the nature of the fire problem for policy makers who must decide on allocations of resources between fire and other pressing problems, and within the fire service to improve codes and code enforcement, training, public fire education, building technology, and other related areas.

The U.S. Fire Administration, which has no regulatory authority, sends an experienced fire inspector into a community after a major incident only after having conferred with the local fire authorities to insure that USFA's assistance and presence would be supportive and would in no way interfere with any review of the incident they are themselves conducting. The intent is not to arrive during the event or even immediately after, but rather after the dust settles, so that a complete and objective review of all the important aspects of the incident can be made. Local authorities review USFA's report while it is in draft. The USFA investigator or team is available to local authorities should they wish to request technical assistance for their own investigation.

This report and its recommendations were developed by USFA staff and by Varley-Campbell & Associates, Inc., Miami and Chicago, and by TriData Corporation, Arlington, Virginia, their staffs and consultants who are under contract to assist the Fire Administration in carrying out the Fire Reports Program.

The U.S. Fire Administration greatly appreciates the cooperation received from Chief Joseph Medina; Deputy Chief-Operations James Olson; Battalion Chief Scott Peoples; and Lawrence Wright, Bureau of Fire Investigation of the San Francisco Fire Department.

For additional copies of this report write to the U.S. Fire Administration, National Fire Data Center, 16825 South Seton Avenue, Emmitsburg, Maryland 21727. For color copies of photographs, see the USFA Web page at http://www.usfa.dhs.gov/

U.S. Fire Administration
Mission Statement

As an entity of the Department of Homeland Security, the mission of the USFA is to reduce life and economic losses due to fire and related emergencies, through leadership, advocacy, coordination, and support. We serve the Nation independently, in coordination with other Federal agencies, and in partnership with fire protection and emergency service communities. With a commitment to excellence, we provide public education, training, technology, and data initiatives.

TABLE OF CONTENTS

Entrapment in Garage Kills One Firefighter

75 Everson Street
San Francisco, California

March 9, 1995

Investigated by: Scott M. Howell

Local Contacts: San Francisco Fire Department
                      260 Golden Gate Avenue
                      San Francisco, CA 94102

                      Joseph A. Medina, Chief of Department

                      James P. Olson, Deputy Chief-Operations

                      Scott Peoples, Battalion Chief

                      Lawrence Wright, Bureau of Fire Investigation

OVERVIEW

One firefighter was killed and eleven others injured, one critically, fighting a residential fire in San Francisco, California, on March 9, 1995. Three fighters were trapped when an overhead garage door closed behind them without warning.

The fire spread rapidly from its origin in a lower level bedroom. Sixty mile per hour winds accelerated the spread of the fire from the rear of the structure towards the front. The fire was initially attacked by entering the structure through the open overhead garage door.

The cause of the door closing is not known. Several of the injuries were due to smoke and hot debris pushed onto the rescuers by the extreme wind conditions as they fought to save the trapped firefighters.

KEY ISSUES

Issues	Comments
Firefighter Fatality	Three firefighters from the first-in company were trapped in the garage. The lieutenant died, while the other two were seriously injured.
Protective equipment and PASS devices	The trapped personnel were all wearing full protective equipment including SCBA and PASS devices.
Rescue	Rescuers had difficulty opening the overhead door to reach the trapped firefighters. Heavy smoke and wind conditions hampered the operation of gasoline powered equipment during the rescue.
Weather conditions	Winds of up to 60 miles per hour from the south and driving rain combined to help cause this fire and greatly accelerate its spread.
Fire cause	The fire was started by arcing or stray current when wind-blown rain entered an exterior electrical receptacle. The heat generated ignited construction paper and the wood siding of the structure.
Fire detection	The house had five smoke detectors. One of the detectors was wired into an alarm monitoring service. The other four were local battery powered detectors. Detector activation alerted the occupants to the fire.
Occupant actions	After being awakened by a smoke detector the occupants searched out the source of the smoke and then attempted to fight the fire before calling the fire department.
Structure and Terrain	The house was built on a steep hillside which made size-up very difficult. From the front only one of the three levels was visible.

THE FIRE BUILDING

This fire occurred at 75 Everson Street, San Francisco, California, in a detached single-family house built in 1965 on the side of a steep hill. The back of this wood framed structure faced south and had large windows to capture the view of Glen Park Canyon.

The house had three levels (see Figures 1a and b). The lower level was only accessible from the back yard and was used exclusively for storage. The middle level had three bedrooms, each with a view to the south. A utility room ran the width of the house on the north side. Two full bathrooms flanked the open center stairway (Figure 3).

The upper floor was the only level visible from the street (Figure 2). On this level, the living room and the master bedroom faced to the south. The garage was north of the master bedroom, while the dining room and kitchen were north of the living room. Between the kitchen and the garage was an entry hall leading to the front door of the residence. The two upper levels contained about 2,050 square feet of living space.

Most of the windows and door of the house were protected by metal security bars or gates. Four foot wide wooden decks spanned the rear of the upper and middle levels outside the bedrooms and the living room. These decks were only accessible from inside the house.

The house was protected by five single station smoke detectors. One of the detectors, which was located at the top of the stairs near the front entry hall, was wired into an alarm monitoring service. The other four were battery powered. One was in the utility room, one in the master bedroom, one in the kitchen and the last in the garage. These four detectors were not monitored and sounded only a local alarm.

The garage door was a single panel, constructed of 3/8 inch plywood with 2x4 and 2x2 framing (Figure 4). This door was 16 feet wide, 7 feet high and weighed about 250 pounds. Connected to the door was a screw drive type mechanical garage door opener. The opener could be activated by a wall-mounted push button near the door leading to the entry hall (Figure 5) or either of two remote controls. The remotes were kept in the family cars.

For Figure 1a, see page 4

For Figure 1b, see page 5

For Figure 2, see page 6

For Figure 3, see page 7

For Figure 4, see page 8

For Figure 5, see page 9

The exposures to the east and west were also single family homes (exposures B & D). Exposure A was the street with additional single-family homes on the opposite side. Due to the slope there were no significant exposures on side C.

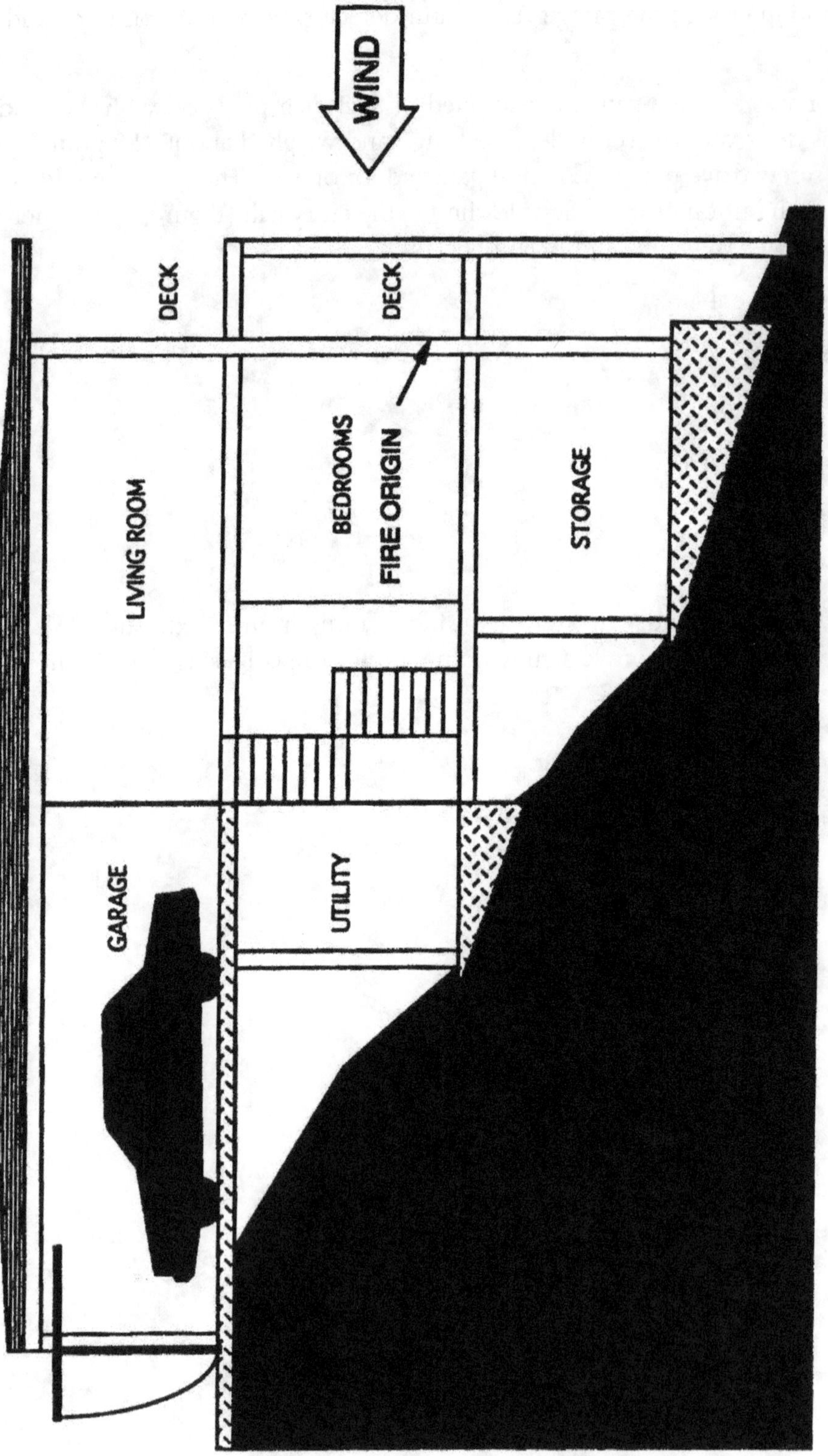

WIND

DECK

DECK

LIVING ROOM

BEDROOMS

FIRE ORIGIN

STORAGE

GARAGE

UTILITY

Figure 1a

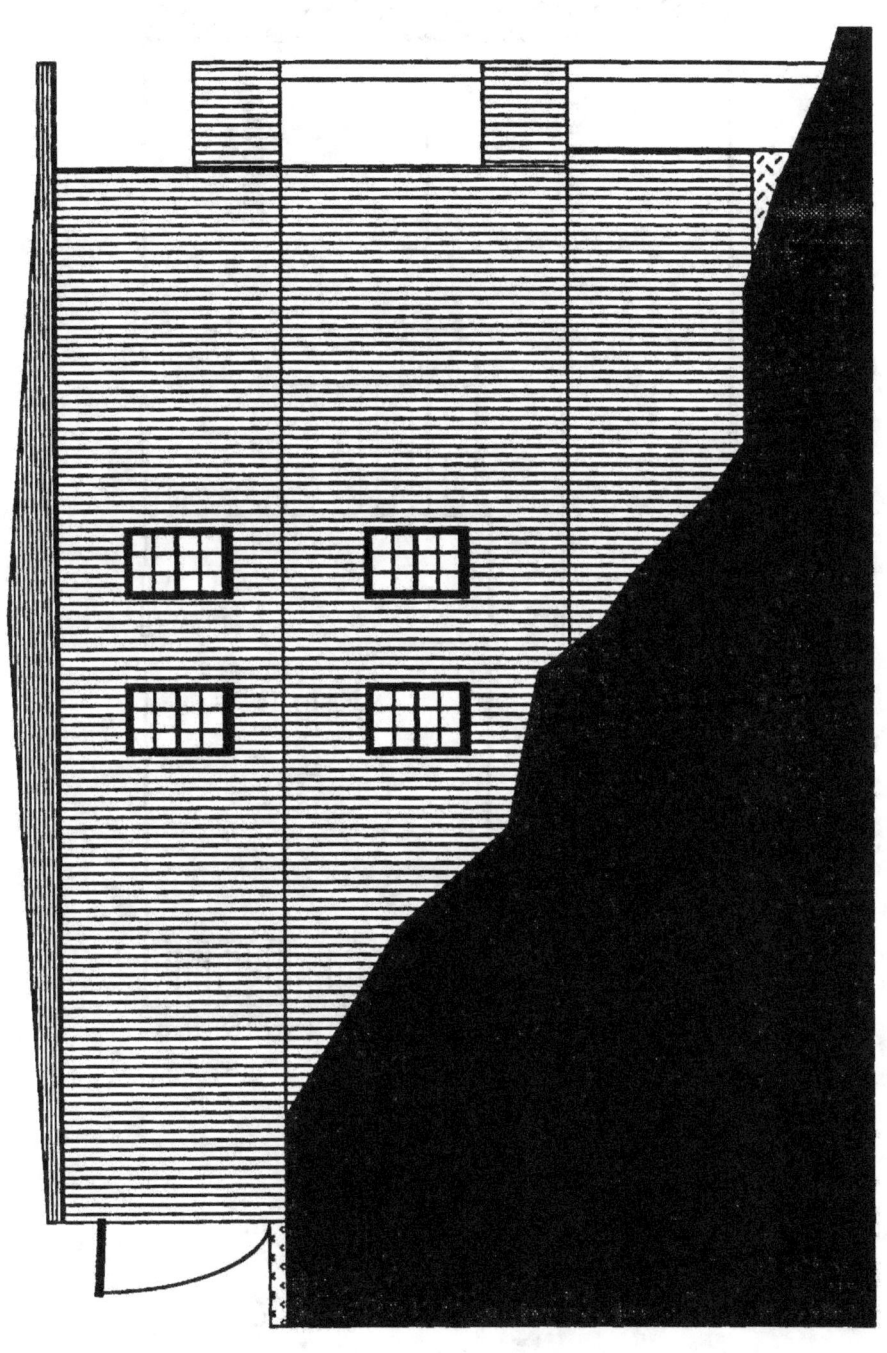

Figure 1b

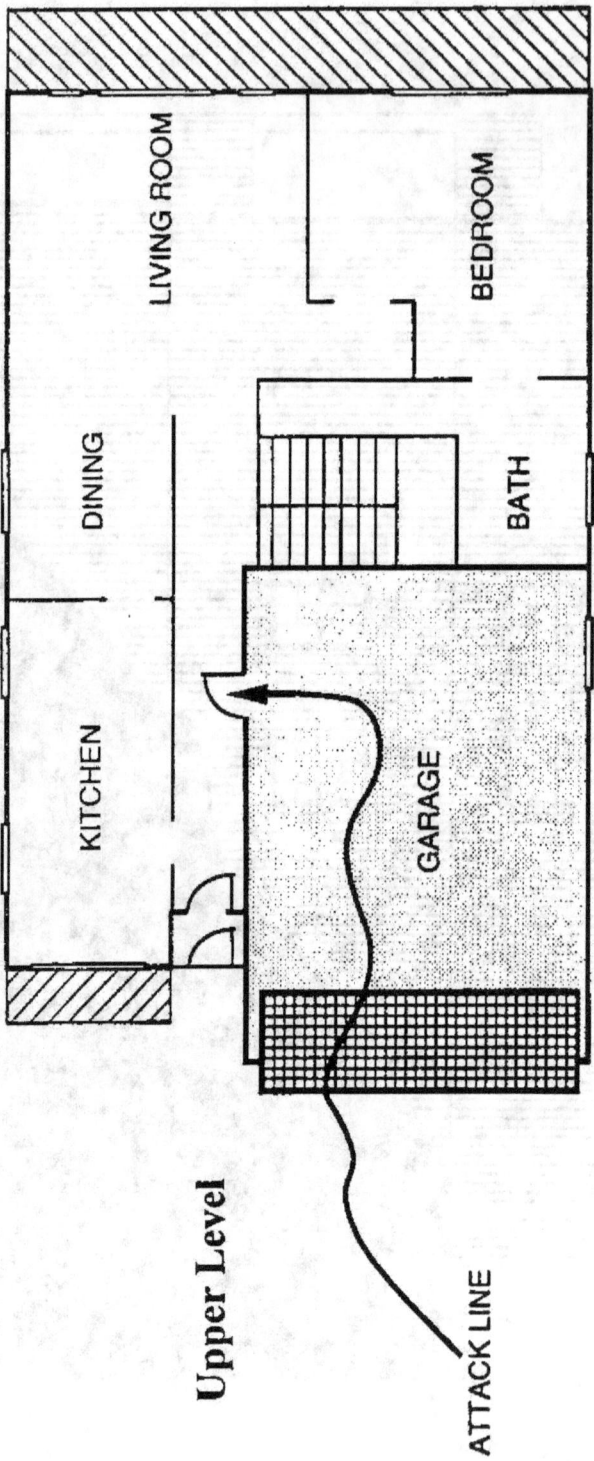

Figure 2

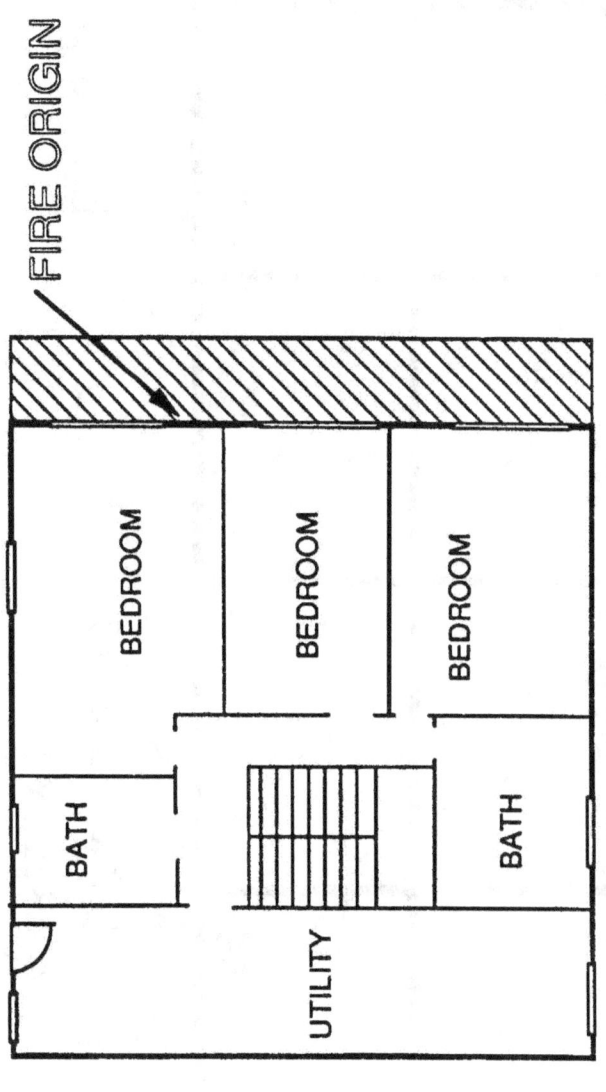

Middle Level

Figure 3

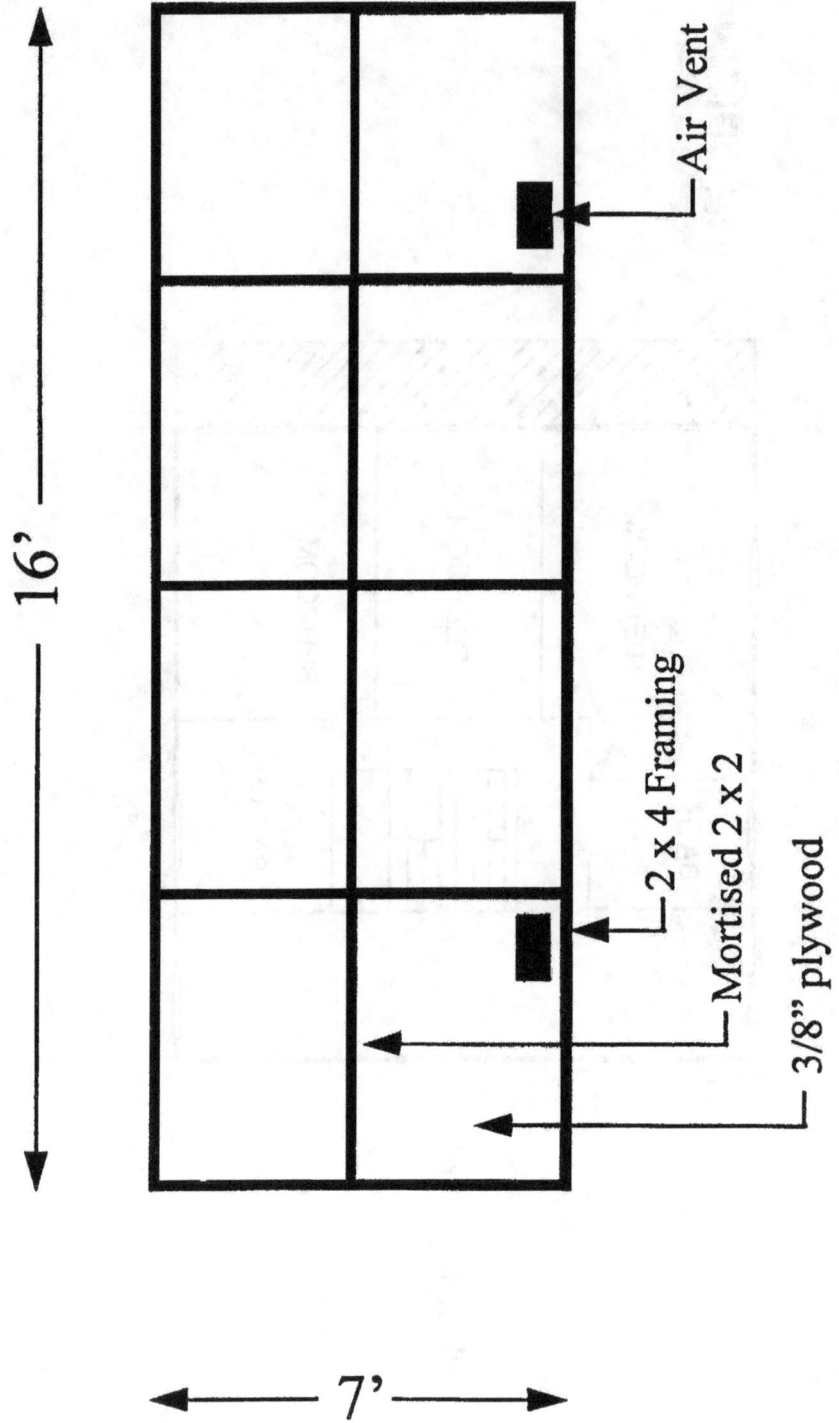

Figure 4

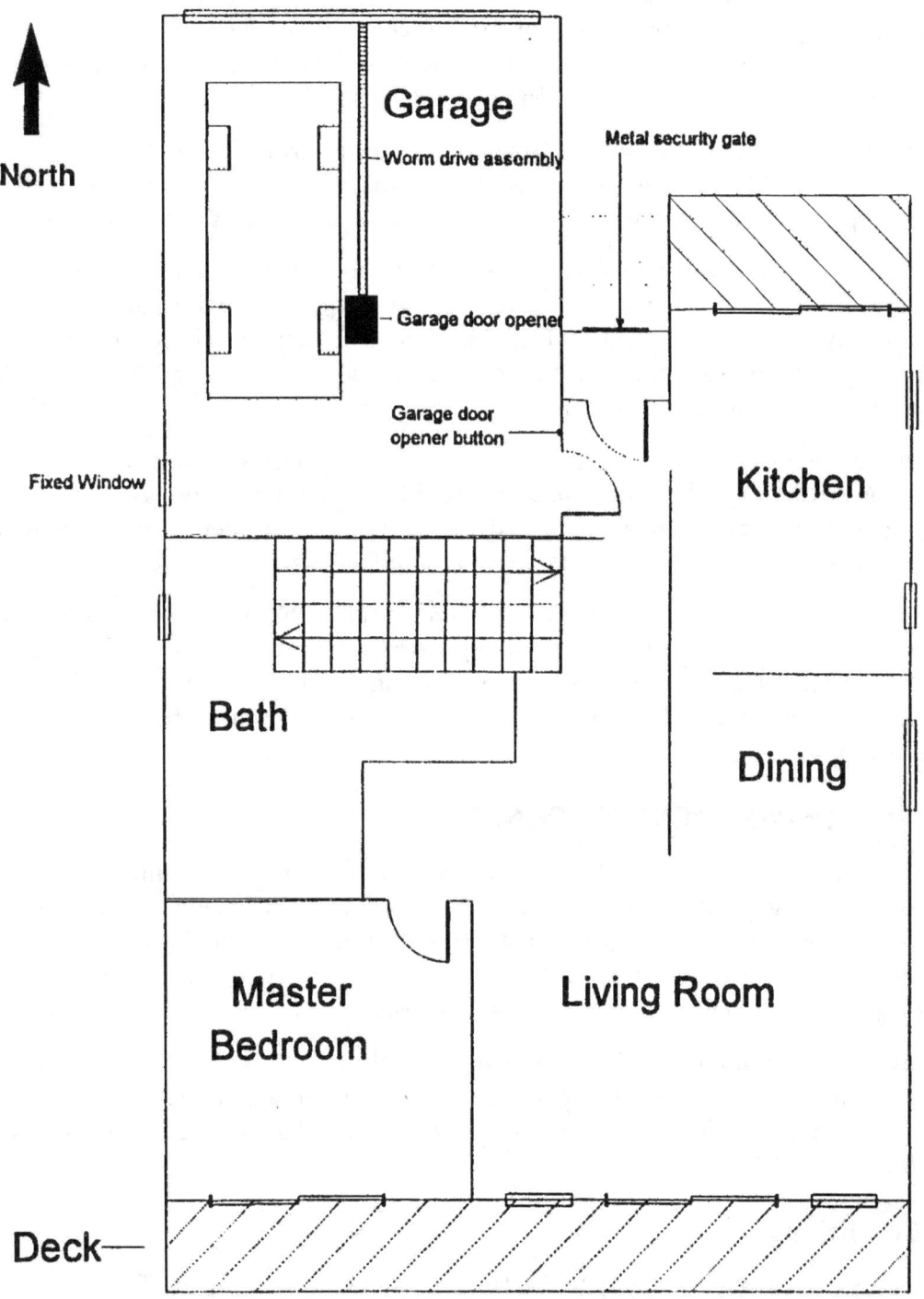

North

Garage

Worm drive assembly

Metal security gate

Garage door opener

Garage door opener button

Kitchen

Fixed Window

Bath

Dining

Master Bedroom

Living Room

Deck—

Figure 5

FIRE DISCOVERY

At the time of the fire two adults and their four children were asleep. The father was awakened at about 12:45 a.m. by the sound of a smoke detector operating. He noticed a light haze of white smoke and spent several minutes looking for the source before he discovered that the smoke was being pushed into the house from the exterior by the wind. The smoke was entering near the bedroom on the middle level where he had been sleeping.

On the outside deck he found a one-foot by one-foot area burning around an outside electrical outlet. He went to the kitchen, filled a pan with water, which he used in an attempt to fight the fire. He refilled the pan twice in the downstairs bathroom, but was unsuccessful in controlling the fire.

At 12:52 a.m. the alarm monitoring service called the home to verify activation of the smoke detector. The person who answered confirmed that there was a fire. Several minutes later the smoke became extremely heavy and the occupant instructed one of his daughters to call the fire department. The family then evacuated the residence. It is believed the bedroom sliding glass door and the utility room doors were left in the open position.

The mother put the two boys into one of the family cars in the garage and backed the car out as SFFD Engine 26 was arriving. The two daughters also exited through the garage. The father was in the garage trying to find a way to remove the remaining car when the lieutenant and a firefighter from Engine 26 approached and asked where the fire was and if everyone was out of the building.

After confirming that everyone was outside, the father went to the sidewalk for a short time, then returned to the garage to retrieve his prescription glasses from the car. Smoke in the garage was not a problem at this time. He remembers seeing the lieutenant and the firefighter, with SCBA on, in the southeast corner of the garage. He then went across the street to watch the fire. Within one minute he saw flames above the roof line of his house.

INITIAL FIRE DEPARTMENT RESPONSE

The alarm monitoring service called the San Francisco Fire Department Communications Center to report an activated residential smoke alarm, but did not report a confirmed fire, at 75 Everson Street at 12:53:30 a.m. Engine 26 was dispatched at 12:54:40 (see SFFD radio log in Appendix B). The crew of Engine 26 included a lieutenant, a 25-year veteran of the department, a pump operator with four years experience, a firefighter with two years experience and a probationary firefighter.

At 12:56:30 the Communications Center received the call from 75 Everson Street stating that the house was on fire. Based on this information a full first alarm assignment was dispatched. In addition to Engine 26, the first alarm assignment consisted of Engines 32 and 11, Truck 15, and Battalion 6.

Arrival Conditions

SFFD Engine 26 arrived on the scene at 12:59:10 and reported a "working fire." The pump operator pulled slightly past the fire building in an effort to give his officer a look at three sides of the scene (Figure 6). A few blocks away Truck 11 had just gone back in service from another call and heard the Engine 26 report. Truck 11 told the Communications Center that they would respond to 75 Everson Street.

The lieutenant and the probationary firefighter, after donning SCBA, entered the garage where they met the father. The senior firefighter followed them, pulling a 1-3/4 inch pre-connected line. The resident told them that everyone was out of the house, the fire was in a downstairs bedroom and it was accessible through the garage. The metal security gate for the door between the garage and the house was open, but the security gate for the front door was locked.

The smoke conditions in the garage were still minor, as evidenced by the resident re-entering the garage to get his prescription glasses from the car.

The battalion chief and chief's aide, BC6 and BC6A, arrived on the scene at 12:59:50. From the front of the house the battalion chief saw fairly heavy black smoke in the garage and flames visible in the front entry to the left of the garage. The aide saw the members of Engine 26 in the garage as they prepared to make an interior attack through the door into the residence.

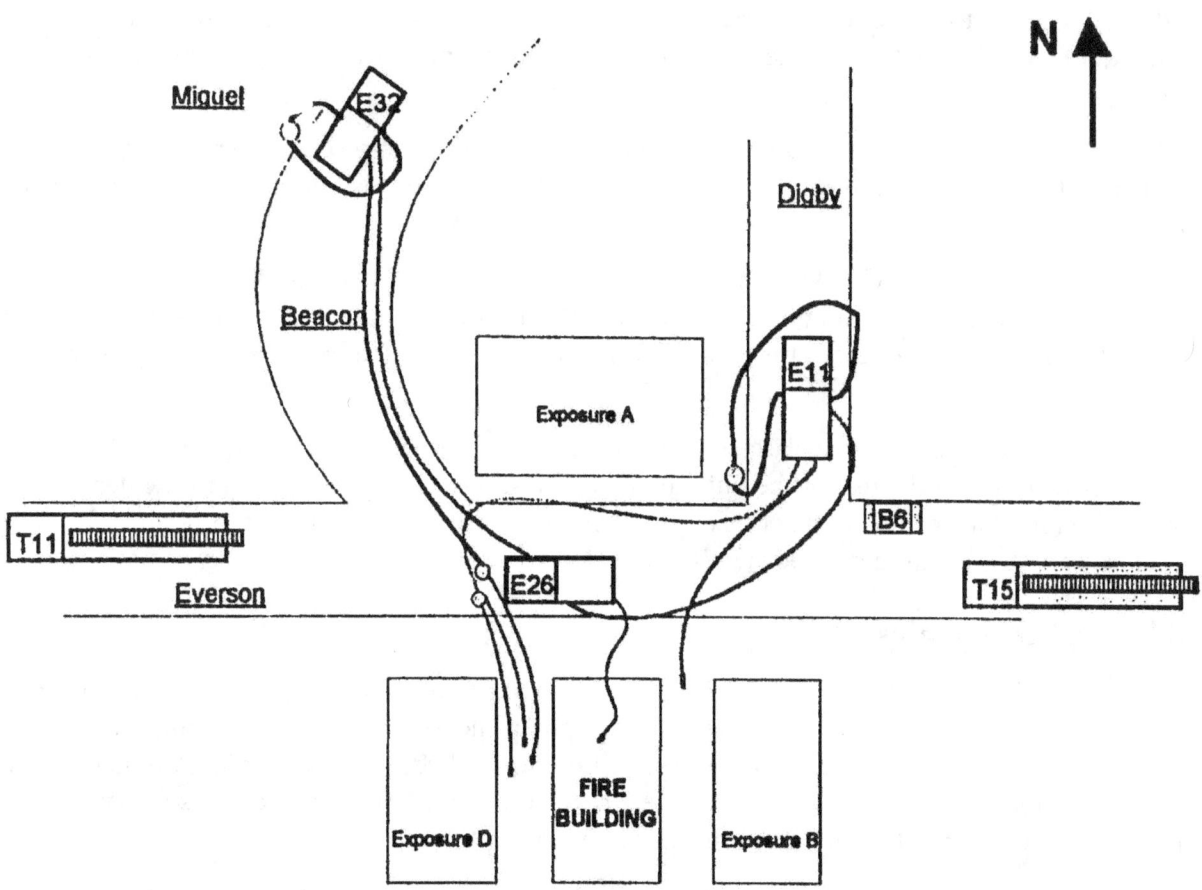

Figure 6

Exposure Problem

The wind was pushing the flames and smoke from the rear toward the front of the house and Exposure B. The battalion chief realized that Exposure B was in jeopardy and ordered his aide to check the rear of the building. The aide found that he could not go between Exposure B and the fire building due to the volume of fire present. He was forced to go around the right (west) side between the fire building and Exposure D, to get to the rear.

From the rear, the aide tried to report to the battalion chief that two floors were fully involved at the rear of the house. His transmission went unacknowledged, possibly due to the heavy radio traffic.

Engine 26 began to attack the fire by advancing the line through the door into the house. The lieutenant opened the door and the senior firefighter deflected the hose stream off the entry hall ceiling for a very short period of time before the wind slammed the door shut. After another similar attempt failed the nozzle was passed to the lieutenant.

A wire shoe rack was used to hold the door open while the lieutenant operated the line into the entry hall for about one minute. In spite of the hoseline, the radiant heat continued to intensify making conditions untenable at the door. Visibility in the garage had decreased to nil when the lieutenant made the decision to withdraw. The senior firefighter followed the east wall, crawling back to the garage door, which was closed when he reached it. Conditions in the garage were deteriorating quickly.

He then moved to the west, but could not find the lieutenant or the probationary firefighter. As he crawled to the west he tried to reach the overhead door release but was burned as he raised his hand about three feet.

The senior firefighter started kicking the door and yelling for help. He could hear, but could not see the lieutenant and the other firefighter yelling for help and banging on the door. He heard a low-air alarm activate and, worried about his own air consumption, attempted to conserve air by remaining calm.

While in the garage, he left three blasts of extremely hot gases. These three blasts are thought to have been from windows in the rear of the building breaking and allowing gusts of wind to intensify the flames. The firefighter estimates that two to three minutes passed between his discovery that the door was closed and the beginning of rescue efforts.

Additional Companies

Engine 32 and Truck 11 arrived on the scene at approximately the same time, 1:00:30 a.m. The crew members donned their SCBA and removed the equipment necessary to set up hoselines to protect Exposure B. Engine 32 then reverse laid two large lines to the hydrant at Beacon and Miguel Streets, a supply line for Engine 26 and a second for exposure protection. The smoke was so heavy at this time that the operator of Engine 32 had trouble negotiating the street.

Two members of Engine 32 pulled a second pre-connected line off Engine 26. While pulling the line they heard yelling from inside the garage and saw that the door was down.

At 1:03:00, a second alarm was ordered by the battalion chief. He reported that 75 Everson was fully involved and there were severe exposure problems. (See Appendix A for greater alarm response). Engine 11 arrived on the scene and was utilized to expand the water supply and lay lines for exposure protection.

RESCUE EFFORTS

From incident analysis and reconstruction conducted by the San Francisco Fire Department, it appears that the members of Engine 26 became trapped at about 1:02:00 a.m., but it was not recognized that they were in trouble until about 1:05:00. At 1:05:50 a third alarm was requested.

In an effort to open the garage door a firefighter pulled on the garage door handle with no result. Five members, including the battalion chief, then attempted to raise the door but there were not enough hand holds available on the bottom of the door to allow sufficient force to be applied.

Two firefighters began to work on the door with axes in an attempt to gain entry. An attempt also was made to gain entry with a Chicago Door Opener, but because of the resilience of the plywood, the door could not be opened. Even with a hoseline cooling the rescuers, they were driven away from the door several times as the products of combustion were blown onto them. The force of the wind was blowing the smoke and flames horizontally towards the rescuers and across the street.

By about 1:09:00 a chain saw and a multi-purpose saw were being used on the door. Both tools would only run for short periods of time before stalling due to the smoky conditions.

The work with the axes paid off at about 1:10:00 when a hole was made in the lower right corner of the door near a vent. The senior firefighter put a hand through the hole in an attempt to escape. The rescuers ripped parts of the plywood off by hand until he was able to crawl out. He informed the rescuers that two others were still trapped inside.

The rescuers then managed to get the gas powered tools to run long enough to cut a hole in the left side of the door. Through this hole the missing personnel could be seen. The door finally failed and the five rescuers were able to raise it at approximately 1:11:00.

The lieutenant and the probationary firefighter were both found face down, just inside the garage door. Their SCBA face pieces were not connected to the regulators. The charged hoseline was in the garage, but not flowing water. The fire had grown so intense by this time that the injured firefighters had to be moved into the street before first aid could be administered. CPR was administered to the lieutenant. The probationary firefighter was initially unconscious and having difficulty breathing. All three injured personnel were transported from the scene by ambulance. The lieutenant was later pronounced dead at the hospital.

FIRE EXTINGUISHMENT

Concurrent with rescue operations, suppression efforts were underway to control the fire. By 1:05 a.m. Everson Street Command was established with Battalion Chief 6 as Incident Commander. Companies responding from the south reported that the rear of the building was fully involved. Staging for the greater alarm companies was set up at Station 26.

The crews of Engines 11 and 32 were able to place two hoselines to the rear and between Exposure B and the fire building. The captain tried to report his progress to the Incident Commander but experienced radio transmission difficulties.

After the trapped crew of Engine 26 had been removed from the garage, Truck 15 was ordered to operate a large handline into the garage. Another handline was put into operation to protect Engine 26 which had become an exposure. Later, Truck 15 cut a hole through the wall of the utility room from the exterior of the building on the Exposure D side. Through the hole they were able to extinguish a large volume of fire in the utility room.

Engine 43, a second alarm company, secured a water supply to the rear of the building, on side C, and with the help of Engine 44, a third alarm company, stretched a large line to the rear of the building to attack the fire.

Division 3 arrived, was briefed and assumed command. Battalion Chief 6 was then assigned as the Operations Chief. Battalion Chiefs 9 and 10 arrived and were assigned to Exposures C and D, respectively. Battalion Chief 2, responding on the third alarm, was assigned to Exposure B.

Engine 24 advanced a large line and a preconnect to the rear of the building. Additional hand lines and a master stream were operated into the front of the house, on side A.

The fire was placed under control at 3:05 a.m. The interior of the dwelling was heavily damaged.

INCIDENT ANALYSIS

The situation that resulted in three members of Engine 26 becoming trapped in the garage can be attributed to a combination of factors. The Lieutenant of Engine 26 was not able to view the rear of the building to make a size-up. There was little indication of the magnitude of the fire from the street when Engine 26 arrived and the occupants were leaving under no apparent distress. These indications understandably may have led the veteran lieutenant into believing the fire was less severe and more controllable than it was. According to several knowledgeable people on the scene, the fire grew with amazing rapidity after the crew of Engine 26 entered the garage. The lieutenant realized that the attack plan could not succeed against the wind. This is evidenced by his order to back out of the garage. The rapid acceleration of the fire can be attributed to the building design, the extensive use of glass opening in the rear of the structure and severe wind conditions.

The most critical factor was the unexpected closing of the garage door behind the attack crew. The reason the door came down is not clear. There are several possibilities listed in the San Francisco Fire Department report on this incident: the door may have come down due to activation by either of the two remote controls or activation of the stationary button mounted in the garage next to the door where Engine 26 was operating. The door may have been lowered due to water or heat interference with the integrity of the circuits of the opener or of the stationary button. The "closer" may have been activated by an outside source, such as a television remote control or an airplane flying overhead.

The possibility of overhead doors closing unexpectedly should be recognized as a potential hazard to firefighters. Some overhead doors are rather heavy and use springs to assist in lifting the weight. With the spring properly tensioned, a 300-pound door can be lifted with 25 pounds of force. A heated spring can lose its tension. Without the spring properties the mechanical advantage is lost and the full weight of the door must be lifted to open it.

Blocking a door open with a sturdy object may prevent it from coming down if the spring is heated. Blocking or damaging the guide tracks may also keep a door in the up position; however, it may not work on a single panel type that swings out at the bottom (as the door in this situation operated).

The presence of a garage door opener adds another dimension to the overhead door problem. If the opener remains energized, the door could open or close at any time if the remote control or pushbutton is activated. Newer doors should not close completely or stay down if there is an obstruction under the door. This has been a Federal Requirement since 1991 and has been a requirement for an Underwriters Laboratory listing since 1973. A charged 1-3/4 inch hoseline should be enough obstruction to cause the opener to reverse and reopen the door.

Exposure to the fireground environment of heat and water could cause abnormal activation of an opener. If the opener is activated abnormally there is no assurance that the safety features of the system will function. These features apparently did not function at the Everson Street fire.

A garage door opener can usually be disconnected from the door with the pull of the release handle. This takes the danger out of activation of the opener. The same result can be obtained by unplugging the opener so it cannot operate. This should not cause the door to close if it is in the full open position, but the door should be blocked open to ensure that it does not close later.

The SFFD report on this fire includes two other incidents in which overhead door openers activated unexpectedly. These are included in Appendix C.

Although none were present at the Everson Street fire, the location of alternate means of egress or a safe refuge zone should always be a consideration when initiating a fire attack. If a secondary egress is not readily available, the primary egress route should be protected at all costs throughout the incident.

At this incident it was very difficult to hear the cries for help over the ambient fireground noise and the roar of the wind. It appears that the heavy smoke, wind, and possible radio problems hindered personnel accountability. It was not obvious from the street, because of the wind pushing the heavy smoke, that the garage door was down. Crews working close to the garage did not immediately recognize that the door being down was a problem. The pump operator of Engine 26 was occupied with securing a water supply. The Incident Commander was working to size-up the fire and develop a strategic plan during the time the door is believed to have closed.

There is no indication that the Lieutenant of Engine 26 attempted to use his portable radio to call for help from inside the garage. The radio was found to be in working order after the fire. At one point the battalion chief's aide attempted to make a report from the rear of the building but was neither heard nor acknowledged by the Incident Commanders. The Captain of Engine 11 had a similar experience when trying to communicate from the Exposure B side. These problems may be attributable to the additional noise created by the wind, temporary radio failure due to wet conditions or elements of the fire building that interfered with the radio signals. Heavy radio traffic may have rendered weak signals from portable radios inaudible.

The reason no signal was received from the interior crew is not known and it still being investigated by the San Francisco Fire Department.

PROTECTIVE EQUIPMENT

The three personnel trapped in the garage were all wearing the full ensemble of personal protective equipment provided to them, including self contained breathing apparatus (SCBA). The protective clothing appears to have been in compliance with current standards and performed within expectations. The clothing showed no indication of failure. Protective hoods, which could have lessened the extent of injuries to the surviving firefighters, were neither provided or used.

All three of the Scott 4.5 SCBA were inspected and tested by the SFFD Breathing Apparatus Supervisor. Two were found with the main cylinder valves fully open. The cylinders on the units used by the lieutenant and probationary firefighter were empty when they were rescued and the bypass valves were open. The SCBA that had been used by the senior firefighter was found with the main valve closed and about 2800 psi remaining.

Differences in air supply duration were illustrated in this fire. All three trapped personnel started using air at about the same time. The SCBA were used a maximum of 12 minutes and two of them were fully depleted. The senior firefighter still had 2800 psi pressure left in his 4500 psi cylinder. He recalls hearing a low air alarm sounding before their entrapment was discovered. He may have been partially shielded from the radiant heat in the garage due to his location behind the car. The lieutenant and the probationary firefighter were unshielded, which may have caused a difference in the stress experienced.

Testing of the SCBA revealed no problems or significant damage that would hamper normal operation of the units. Each had been visually inspected and operationally tested on March 8 at the beginning of the shift, according to the pump operator. Each SCBA had a full cylinder at the beginning of the shift.

Each SCBA was equipped with a Personal Alert Safety System. The PASS devices were in working order when inspected at the beginning of the shift and when tested after the fire. There are no reports indicating that the PASS devices operated at the scene or if they were in the "on" position at the time of the rescue.

LESSONS LEARNED AND REINFORCED

1. **Overhead doors require special attention.**

   This incident illustrates the danger of overhead doors to fire personnel. Overhead doors, including fire doors are generally very heavy. Once closed they may been impossible to reopen under fire conditions. If entry is made through an overhead door, steps must be taken to ensure that the door will not close.

2. **Getting out of a fire safely is more important than getting in.**

   Companies entering a building should always seek secondary exits. These exists must be maintained at all costs. A secondary exit at the Everson Street fire was obstructed and not visible because it was blocked by a vehicle. This is often overlooked in residential fires because the risk is not recognized. In many situations a hoseline will help protect the exit path and keep a swinging door from closing. Windows are often acceptable as emergency exists for firefighters. In some situations the secondary exit points must be guarded, even at the expense of slowing the fire attack, to ensure firefighter safety.

3. **Extreme weather conditions require re-assessment of routine response levels.**

   Extreme wind conditions can speed the spread of fire and quickly turn a routine fire into a critical situation. Several of the personnel interviewed during the SFFD investigation of this fire commented on the extremely rapid spread and intensity of the fire. Any weather condition out of the ordinary is a signal that routine response levels should be evaluated for appropriateness. The effect of weather on strategy and tactics should always be a concern. At this incident, the Incident Commander called additional alarms very quickly, in an effort to gather the resources necessary to fight the fire and attempt to rescue the firefighters.

4. **A thirty-minute SCBA will not deliver thirty minutes of breathing air.**

   There are many factors that affect SCBA duration. Both physical and emotional stress cause an increase in the consumption of oxygen and therefore air. A person's physical size and conditioning are also major factors in air supply duration. Each firefighter should know how they react to stress with SCBA on. These reactions affect the duration of the air supply. A certain level of stress can be relieved if training is sufficient to make firefighters comfortable and confident with their equipment.

5. **Delayed reporting can greatly increase the potential for loss of life and property damage.**

   The activation of a smoke detector alerted the residents to a potential problem. If the residents had called for help as soon as they discovered the fire, Engine 26 may have arrived on the scene when the fire was still in its incipient stage. There was a delay of several minutes, even after the fire was located, before the fire department was notified. If the alarm company had relayed specific information about the presence of a confirmed fire, rather than only an activated detector a greater response would have been initially dispatched.

6. **Protective equipment is only effective if it is used.**

Eleven firefighters were injured at this incident. As many as five of the injuries could have been prevented if the firefighters had worn all of the protective equipment supplied to them. Several of these injuries occurred during the rescue efforts; however, failing to use SCBA and other protective equipment in these situations is likely to put the rescuer in need of rescue.

The trapped firefighters were all wearing PASS devices. Information regarding the position of the activation switches of the PASS devices was not available.

7. **Importance of a 360 degree size-up.**

The dangerous size of this fire was not apparent from the front of the building. A complete size-up should include a view from every possible side of the incident scene. If decisions have to be made without all the facts, every effort should be made to gain additional information and ensure the safety of crews until a complete size-up can be accomplished.

8. **Use natural forces as advantages whenever possible.**

The fire service long ago began taking advantage of the natural forces in attacking fire. Vertical ventilation, making a hole in the roof, uses the buoyant properties of heated gases to move the products of combustion out of an enclosed space. In extreme wind conditions, changes in tactics should be considered. Attacking from the windward side allows personnel to operate in a safer area. The convective heat of the fire, as well as the toxic products of combustion, are transported with the wind. Hose streams carry to their target and the steam generated moves through the fire on the wind cooling the scene. Fire streams operated against the wind are often ineffective and may be converted to steam that is pushed back onto the crew by the wind. This may not have been a viable alternative in the Everson Street fire due to the topography of the scene.

APPENDICES

A. SECOND AND THIRD ALARM RESPONSES

B. SFFD COMMUNICATIONS CENTER RADIO LOG

C. "EXPECT THE UNEXPECTED," ARTICLE BY LORNE
    ULLEY, FIELD CORRESPONDENT, HAMPSTEAD,
    QUEBEC, FIRE FIGHTING IN CANADA, APRIL
    1995.

D. PHOTOGRAPHS

APPENDIX A

SECOND AND THIRD ALARM RESPONSES

SECOND AND THIRD ALARM RESPONSES

Second Alarm Response

Engine 24 Truck 7

Engine 7 Division 3

Engine 43 Battalion 9

Engine 9 Battalion 10

Rescue 2 Arson Squad 4710

Service Squad 1 Bureau of Equipment

Third Alarm Response

Engine 33 Truck 9

Engine 42 Truck 6

Engine 44 Battalion 2

Engine 39 Chief of Operations

Chief of Department

APPENDIX B

SFFD COMMUNICATIONS CENTER RADIO LOG

SFFD COMMUNICATIONS CENTER RADIO LOG

12:53:30 a.m.	SFFD Communications Center receives phone call from alarm monitoring company reporting an activated residential smoke detector.
12:54:30	Engine 26 dispatched to investigate an activated residential smoke detector.
12:56:30	SFFD receives 9-1-1 call from 75 Everson stating that the house is on fire and that the fire is downstairs.
12:57:30	First alarm assignment dispatched for a reported structure fire, E32, E11, T15, BC6 dispatched.
12:59:10	Engine 26 reports on the scene with a "working fire."
12:59:30	Truck 11 reports in service from a previous assignment. They are within blocks of Everson and Digby and inform the Communications Center that they will respond to 75 Everson.
12:59:50	Battalion 6 reports on the scene.
01:00:30 *(Approximate time)*	Engine 32 reports on the scene. Truck 11 is already on the scene.
01:00:50	BC6 notifies Communications Center that Truck 11 will be held at this incident.
01:02:10	9-1-1 call from 67 Everson (Exposure B) saying that the back of the house next door is on fire.
01:03:00	BC6 reports that the structure is fully involved and there are exposure problems. A second alarm is requested.
01:03:20	Second alarm dispatched, E24, E7, E43, E9, T7, D3, B9, B10, RS2, SS1, 4710, and BOE. At this time BC6A can be heard trying to report to BC6 that the rear of the building is fully involved.
01:04:30	Truck 15 reports that they are responding and they can see that the rear of the building is fully involved.
01:04:50	Everson Street Command established by BC6 and reports heavy wind conditions.
01:05:00	BC6 reports exposure problems.
01:05:50	Everson Street Command requests a third alarm.
01:06:30	Third alarm is dispatched assigning E33, E42, E44, E39, T9, T6, B2, CD1 and CD2 to the incident.
01:07:10	Staging area set up at Station 26.

01:09:15	Truck 15 and Engine 11 on the scene.
01:10:30	Communications contacts C-MED dispatch center and directs an ambulance and supervisor to respond to the greater alarm.
01:11:50	BC6A requests an ambulance.
01:14:30	BC6A requests two ambulances and reports firefighters down.
01:18:00	Service Squad 1 on the scene.
01:20:30	Everson Street Command reports two firefighters down and CPR in progress on one of them.
01:21:20	4710 (Fire Investigation) on the scene.
01:21:20	CD1 (Fire Chief Medina) on the scene.
01:28:30	Everson Street Command reports the fire is out of control and heavy winds are hampering operations.
03:05:00	Everson Street Command reports fire under control.

APPENDIX C

EXPECT THE UNEXPECTED

EXPECT THE UNEXPECTED

by LORNE ULLEY,
Field Correspondent, Hampstead, Quebec
Fire Fighting in Canada, April 1995

It's no secret to firefighters that their job is a dangerous one, having inherent risks which certainly can involve the likelihood of being injured during an incident.

Cellar fires have always posed a challenge to the firefighter; today there are not just cellars for storage but playrooms or recreation rooms, usually with a heavy fire load along with the furnace room, laundry room and workshop area. Most of these areas have only one way in and out.

So it was in the early hours of February 24, 1994 at 03:27 that the Hampstead (Que.) Fire Department received a call for a fire in a private home with smoke coming from the cellar.

The owner had arrived home after closing up his restaurant and upon opening the garage door he encountered heavy smoke; he closed the garage and went to a public phone to call 9-1-1.

Arriving at the scene at 03:32 the officer-in-charge, Platoon Chief Richard Anderson (subsequently appointed Director) did his size-up: they had heavy fire in the basement of an unattached split level home, made of brick and wood with a garage. Using an automatic garage door opener, the owner activated it to allow the firefighters entry.

Anderson had Lt. Dubeau (now Platoon Chief) and Firefighter Laplante advance an 1-3/4 inch line into the basement, down the stairway from the garage. At this time he called for two mutual aid departments at 03:34 from Cote Saint-Luc and Westmount.

The fire was in the basement recreation room which had finished walls of veneer plywood. The fire extended up the wall, crossing over the ceiling, emitting off heavy smoke; heat conditions were intense as the fire had also entered the furnace room, setting ablaze a work bench.

Mutual aid crews were given their assignments upon arrival. Cote Saint-Luc had its crew relieve the interior attack crew, while Westmount was to carry out ventilation, setting up a PPV fan and checking for the fire extension into the ground floor of the home.

The owner reported that the house was empty so there were not citizens' lives at risk. A routine search was carried out in the event the information was not correct, but no one was found.

In the meantime, at 03:53, two more mutual aid departments were called. Outremont and the Town of Mount Royal. It was at this time that Hampstead's Dubeau and Laplante were relieved by the Cote-Saint-Luc crew who took over the line, continuing the interior attack.

The Hampstead lieutenant led the way out from the recreation room to the garage. Once in the garage, they followed the hoseline, but to their surprise the door was now closed and they were low on air – and they had no idea where the interior button was located to activate the automatic doors.

Just prior to this, Lt. Earl Graham and his personnel from Westmount saw that the double garage door was closing. What seemed to be only seconds after the door had closed they heard banging from the inside. Unknown to them at the time, it was the interior attack officer. He had run out of air, taken off his SCBA and was lying on the floor trying to get fresh air from the crack at the bottom of the door.

They called the owner over to use his door opener which he did while the Westmount crew heard their fellow firefighter in trouble. Once the door opened, they assisted him to the outside. What could have been a more serious situation with possible injuries was avoided.

It was later determined that one of the members of the relief crew going in had brushed against the door switch causing the garage door to close, unbeknownst to them.

Firefighters, after learning of this potentially serious accident, wished to share this experience with fellow firefighters.

It is important to follow some very minor rules taught during training, in this case to always place something under an overhead garage door in the event it does close for some reason. Use a pike pole or ladder or even an item found at the structure involved in the incident. At this incident, the chief officer saw to it a ladder was placed under the door once fully opened following the exit to the firefighters.

Cause of the outbreak was determined to be a ceramic heater that set a chesterfield (sofa) in the basement playroom on fire.

The Hampstead department has a policy of not cutting the power until it is found to be necessary, using the power to their advantage for lighting, etc.

Weather conditions at the time of the fire: -12 degrees C, with snow.

APPENDIX D

PHOTOGRAPHS

PHOTOGRAPHS

All photographs were provided by the San Francisco Fire Department Bureau of Fire Investigation.

PHOTOGRAPH 1: The front of the house. The garage is to the right, the entryway door is in the center and the kitchen is to the left side of the photograph.

PHOTOGRAPH 2: The rear of the house as viewed from ground level. The fire started in the wall above the lower deck to the right center of the photograph.

PHOTOGRAPH 3: The northeast corner of the house. Note the security gates on the lower level openings.

PHOTOGRAPH 4: The southwest corner of the house. Note the steepness of the building lot and the proximity of the house next door.

PHOTOGRAPH 1: The front of the house. The garage is to the right, the entryway door is in the center and the kitchen is to the left side of the photograph.

PHOTOGRAPH 2: The rear of the house as viewed from ground level. The fire started in the wall above the lower deck to the right center of the photograph.

PHOTOGRAPH 3: The northeast corner of the house. Note the security gates on the lower level openings.

PHOTOGRAPH 4: The southwest corner of the house. Note the steepness of the building lot and the proximity of the house next door.

www.ingramcontent.com/pod-product-compliance
Lightning Source LLC
Chambersburg PA
CBHW081237170526
45165CB00009B/3082